I0511369

Vivre A Jamais

Why We Go to War

Emilio Vick Garcia

Copyright 2015 Emilio V. Garcia

All rights reserved. No part of this publication may be reproduced, stored in a retrieval system, or transmitted in any form or by any means—electronic, mechanical, recorded, photocopied, or otherwise—without the prior permission of the copyright owner, except by a reviewer who may quote brief passages in review.

Library of Congress Control Number:**1-2082554861**

Small groups of primitive prehumans did not go to war, did not use writing as we who are civilized, and did not store provisions for leaner times. They lived as hunters but not as just carnivores. The men would go out to hunt, and the women would collect edible fruits, vegetables, seeds, nuts, and insects. This primitive prehuman animal took from nature whatever was edible through trial and error and used animal skins as clothing.

They did not plan attacks on each other, but if a stranger wandered into their territory, he or she would be killed and eaten. Cannibalism was a part of all primitive people's diets; there were no funerals. And, after lightning struck a primitive prehuman, they discovered it was easier to eat someone if he or she was cooked first.

Fire became a god.

With fire, the prehuman cooked plants, animals, other prehumans, dogs, and birds whatever was edible.

With fire, food became softer to eat—and so began the degeneration of human teeth.

Some may say that all animals have evolved. Then why are humans the only animals that have conceptual thought, and why are other animals not talking to us?

No. Humans did not evolve and become what they are today—they were affected and became what they are today. I'll explain as we move on in this book.

Humans invented tools. They found sticks, and the sticks became weapons. They domesticated animals and used them for food and milk. They planted crops.

They built huts—houses of all sorts—and out of this emerged towns, cities, and civilization. With the invention of the wheel came trade, as people transported products, animals, vegetables, fruits, and humans. To trade as food.

From hunting to agriculture to ownership, along with civilization and private property, came slavery, which has created industry in all existing societies. Having slaves was a good thing.

Slaves did the work no one wanted to do, and owning them was profitable. The state emerged for the protection of property, to regulate the classes, and to wage war. The purposes of war are generally to plunder, rape, enslaved, and exploit a society's neighbor, but some wars are of morals. A dictator's reaction to its neighbor's arrogance.

The great Persian Empire, with Darius as its leader, set out in 400 BC with a small force to teach Athens and Eretria, what Darius considered petty states, to mind their own business. After destroying Eretria, Darius turned to Athens.

Not wanting to directly attack a walled city with a standing army, he ended up paying a heavy price. Darius landed at

Marathon, twenty-four miles away from Athens, with a strategy—to draw the Athenian army out of the city and, in their absence, attack the city.

It would have been a good plan, if not for Miltiades and his Greeks. With their superior armor and longer spears, leaving the city of Athens, they won the victory at Marathon.

Rush back to Athens in time. To defend the city.

The Persians, given no choice, sailed back to Asia.

Move forward to 480 BC; the Persians had built an army too big to move by sea and instead moved overland, with their navy supplying the army. They moved along the coast, with no choice but to take a direct approach. This direct approach gave the Greeks opportunities to attack the advancing Persians at key points along the road, but it was the sea battle at Salamis that was the deciding factor.

The Athenian General Themistocles with his

Greek War galleys lured the Persian galleys into the narrow

straits; the Persians became congested, and the Greek War

galleys attacked their flanks. The Persian army could do nothing

but watch the destruction of their fleet and their only source of

supplies. This was the end of the second invasion of the

Persians.

At the same time King Leonidas and his 300 Spartans made

History

After all of this the city states of Greece were back to fighting

with each other until

Philip of Macedon's indirect approach of taking his army

through Cytinium and Elatea and, by surprise, descending on the

Athenian army, he turned a new page in history and began

Macedonian supremacy in Greece.

Alexander inherited a plan of conquests, an army, and a grand strategy. His superior strategy and tactics led him to outmaneuver the Persians when confronted by the last Persian army. During the conflict, Alexander distributed his army widely and made noisy marches to confuse his opponent. At night, with a small force, he crossed upstream and surprised the Persian army. Alexander's masterpiece of indirect tactics unbalanced his enemy and won a victory against the Persian army.

Hannibal's indirect war strategy and overland invasion thus the battle at the plains of Lombardy, where his superior cavalry claimed victories at Trebia and Ticinus. Hannibal did not take the normal roads to Rome, instead marching his army through the marshes. Although he lost many men, this tactic ultimately allowed his army to march past the Roman army undetected—and then Hannibal devastated the country.

Soon the Roman army was in pursuit, but Hannibal set a trap and, by surprise, annihilated the Roman army. Hannibal did not directly attack Rome, a fortified city with a standing army. He knew this would give Rome an opportunity to win the war.

Still, Hannibal's indirect war strategy was challenged by the Roman general Fabius. his success, famous in history, Fabian, invented guerilla warfare.

The Roman general, Publius Scipio the younger, was sent to Spain. There, with superior tactics of an indirect attack on Hannibal's base, he was able to deprive Hannibal of his reinforcements and supplies.

In 204 BC, Scipio landed in Africa, lured Carthage's cavalry into a trap, and destroyed it. Syphax marched an army of fifty thousand men to reinforce Hasdrubal Gisco and the Carthaginian army. The combined armies were superior to Scipio's army.

Scipio fell back to a fortified small peninsula and distracted the enemy with false preparations for a seaborne attack against Utica. Instead, under the cover of night, Scipio's army attacked the enemy's two camps.

He first set fire to Syphax's huts, and in the confusion the Romans penetrated into the camp. Imagining the fire to be accidental, Hasdrubal's Carthaginians opened their gates. Scipio raced forward and attacked them, making a breach, and both hostile armies were routed.

Instead of a direct attack on Carthage, Scipio cut off the city's supplies and allies.

Bypass the city, then Scipio marched his army into the interior of Carthage away from the city and its main source of supplies. Hannibal in pursuit found his army far from the city of Carthage without water or reinforcements.

On Hannibal's arrival, Scipio fell back to a plain, where Scipio's cavalry defeated Hannibal's cavalry and annihilated his army. This was followed by a bloodless city of Carthage surrender, and Rome became the dominant power in the civilized world.

It is amazing, and a bit weird, that democracies failed to anticipate Hitler's objectives after reading *Mein Kampf* and his speeches.

Today, Iran comes to mind; their objective is to destroy America—it is in their speeches and the actions they are pursuing. In addition, we are about to give them all the supplies one hundred billion dollars can buy.

Can we not see what is obvious? They—Sayyid Ali Khamenei and Hitler—share a common goal: their conception of people's right to dominate.

Looking to build a totalitarian state filled with people to be bred for slaughter, Hitler employed an indirect approach that was diplomatic, militaristic, and psychological; all of his attacks had alternative objectives.

He gained control of Germany by using a psychological strategy of socialist and capitalist interests. With control of Germany, Hitler negotiated a ten-year peace pact with Poland, covering his flank. Hitler then indirectly approached the rear of Britain and France by supporting General Franco to overthrow the Spanish republican government.

With peace marches and propaganda and the penetration of psychological ideas, Hitler was a master strategist. But in grand strategy, he failed. He allowed the British forces to escape at Dunkirk. He changed from an indirect approach to a direct approach. And conquering a people without securing peace produced a misery in the conquered people that turned into widespread resentment and resistance.

Hitler failed to make peace with Britain after his victory on the continent. He failed to win the war with Russia. And the balance of power turned against him.

This tells us in War the best way humans have killed humans, is the indirect approach but it does not tell us what the cause is or if it has a purpose.

The real question, then, is this: What is it that can change prehuman behaviors and give humans conceptual thought?

Can we find the answers with the great philosophers?

Socrates did not write a word. We have his legacy through an interpretation by Plato, a student of his. Plato wrote that true wisdom consisted in knowing that you know nothing. He expanded on what Socrates had to say—delving into topics such as knowledge in shadows, optical illusions, opinions, different levels of knowledge, and aspirations to a higher level. Plato's point was that perception—or what we get from our senses—is not true knowledge.

Aristotle agreed with Plato, except Aristotle put them together as a formed matter for whatever purpose or end it served.

Descartes went on to say that it is all about reason, the most important element in human knowledge. Descartes, the rationalist, let mathematics solve the problem—my consciousness is all that can be known.

Then there was Hume the empiricist. From him we get two questions: How do you know? What are the limits of knowledge? In essence, do you have any scientific experiments that prove your philosophic worldview? No? Then it is all rubbish.

The philosophes—Voltaire, Diderot, we need not name them all—did not debate rationalism versus empiricism; they used whatever worked for their political purposes to create the French revolution, looking for the day when men would be free from tyrants and the church. That turned into mob violence and the rise of Napoleon.

Kant's answer to the empiricist was that there are structures in our minds and our consciousness, that laws and nature hold true. Our minds structure them.

Hegel's theory of dialectic outlined the three stages. The triadic, the thesis, was the first stage. The antithesis, that which contradicted the first, was the second. And the synthesis, the third stage, was a new stage that emerged from the first two.

The owl of Minerva spread its wings and took flight only when the shades of night were falling. Minerva was a symbol of the Roman goddess of wisdom.

What all of this means is?

The spirit of the people, whatever existed in the status quo, was absolute when culture had matured. It was too late for society to change.

Hegel set the groundwork for Marxism and the *Communist Manifesto*.

Different from Hegel's phenomenology of spirit, Marx's view was that material production created history and that human thought, culture, and society lay in its material economic foundation.

The conditions of production.

The forces of production.

The relations of production.

The division of labor.

Class struggle.

Economic foundation.

The mode of production in material life determined the general character of the social, political, and spiritual processes of life; it was not the consciousness of humans that determined their existence—on the contrary, their social existence determined their consciousness.

Friedrich Nietzsche.

And his Superman philosophy

It is possible to trace all instincts of an animal to the will to power—an inner will and an insatiable desire to manifest power or the application and exercise of power as a creative instinct. Not mankind, but superman, is the goal.

No wonder Hitler loved this person.

The substance of philosophical jargon is, to me, mental masturbation.

But let us get back to pre humans and primitive man. They roamed the planet, surely, as a herd. Banding together, barely walking upright, the herd experienced a mental regression, which resulted in the emergence of a leader and the willingness of the herd to follow its leader, even to its own death. Hunting together, they were susceptible to all things in their environment—predators, bacteria, viruses, climate, and an innate driving force to survive.

Can bacteria and viruses change animal behavior? Some bacteria use humans to promote their own lives. There are friendly bacteria in human stomachs that they cannot live without.

There are multitudes of microbes, viruses, and parasites that use animals, insects, and the environment to travel from one host to another to perpetuate their own survival.

To achieve this, all parasites alter the behaviors of their hosts. The rabies virus makes it difficult for the infected animal to swallow; the foaming of the mouth contains the rabies-infested virus. Then the virus infects the animal's brain and changes the animal's behavior, inciting an aggressive mental state in which they bite. And hence the virus moves on to another victim.

Human papillomavirus (HPV) causes cancer. This virus colonizes the oral cavity and

Changes the infected human's behavior. It makes people kiss more passionately.

The single-celled parasite called Toxoplasma gondii causes infected women to be more outgoing, friendlier, more promiscuous, and more attractive to men. Infected men are anti-social, less attractive to women, and suspicious.

Malaria (protozoan plasmodium) makes infected humans sick, immobile, and attractive to mosquitoes that carry and spread the disease.

The gypsy moth caterpillar, after getting infected with a virus (baculovirus), climbs to the top of a tree, dies there, and becomes a liquefied bag filled with the virus. When it rains the virus falls all over the tree's leaves, and new hosts eat the leaves and are infected.

A specific spider (plesiometa argyra) builds a web like a bull's-eye, which attracts a specific wasp. The wasp (hymenoepimecis Argyraphaga) stings the spider. The spider falls asleep, and the wasp lays an egg on the spider's abdomen. The egg hatches, and the larva feed on the spider's blood. When the larva is ready to build its cocoon, it injects the spider with chemicals. The spider in turn builds a web just for the larva and then sits in the web and does not move. The larva eats the spider before building its cocoon in the reinforced web that the spider built. And, in time, an adult wasp emerges.

The common cold, the flu virus, will not kill you, but it will make you sneeze, cough, and mobile enough to spread the virus.

Ultimately, host manipulation can change the victim into a different species. In humans, retroviruses do just that. Retroviruses are made of ribonucleic acid (RNA) with a protein coat. They are not thought of as alive because they can only replicate inside the cells of another organism. They are a piece of code that is capable of copying itself through a process and using an enzyme called transcriptase.

Normally genetic code is transferred from DNA to RNA; the HIV virus actually reverses the information flow and literally can change a person's DNA.

The idea persists that a human's genetic code can change dramatically by the influence of environmental stress. Think of the human genome as you would think of the immune system—let's call it the genetic genome system—where the immune system is in constant change, and any change is provoked by environmental stress.

It has been discovered that jumping genes called transposons are active under times of stress, jumping and inserting themselves into other genes and changing them. They pick the part of the genome that makes a change that will be beneficial to the animal or that will solve a problem to prevent illness.

Consider the radical Muslims after generations of frustrations. Their genetic genome systems created a violent gene that vents their anger. They cannot control their behavior and are easily moved to immoral, abnormal, inconceivable, violent behavior.

Consider American black slaves after generations of frustrations. Their genetic genome systems created an anger gene that vents their frustrations. They cannot control their anger and are easily moved to criminal behavior.

We can move across the planet and find many examples of behavior that was created by the human genetic genome system because of environmental stress.

Barbara McClintock received a Nobel Prize in 1983 for the discovery of the jumping genes. Jumping genes use reverse transcriptase to change the DNA of the host gene, just like retroviruses.

A virus is made up of a core of genetic code. Ribonucleic acid a polymeric molecule implicated in various biological roles in coding decoding regulation and expression of genes. In addition, a protective coat called capsid, which is made up of protein.

A retrovirus ("retro" means "backward") targets a host cell; once inside the host cell's cytoplasm, the virus uses its own reverse transcriptase enzyme to insert genetic code from its RNA to the DNA of the host. The action of reverse transcriptase makes it possible for genetic code from a retrovirus to become permanently incorporated into the DNA genome of an infected cell.

So-called endogenous retroviruses (ERVs) are persistent features of the genomes of many animals. ERVs consist of genetic code of extinct or fossil viruses, the genomic constitution of which is similar to that of extant retroviruses.

Human ERVs (HERVs) have become distributed within human DNA over the course of evolution; they are passed from one generation to the next and make up 5 percent of the human genome.

Prehuman men and women were exposed to an ancient friendly retrovirus; it's now environmentally extinct but a part of every living human—like the friendly bacteria in your stomach. This retrovirus inserted a multitude of genetic code into the DNA of the prehuman animal, changing this creature into a human with conceptual thought to create language to invent culture to create art to invent war. This ancient friendly endogenous retrovirus is thriving in our DNA, and we are better because of it.

These endogenous retroviruses have placed genetic code in humans that ensure we do not overpopulate the planet—turning hormones off in men around forty-five to fifty years old, beginning the aging process, starting menopause in women.

We are not Gods

We are not a special entity

We are just an animals infected by an ancient, friendly parasitic, retrovirus that has changed us into thinking animals, that has given us conceptual thought

We are a product of an ancient friendly retrovirus that has inserted in to our genome.

Genetic code that is a part of our DNA giving us much more then conceptual thought it has given us the ability to dominate planet earth

To kill each other in war, and to die of old age.

All to perpetuate it's own survival.

Let us name this ancient, friendly endogenous retrovirus that is a part of our DNA "Emilio's friendly human endogenous retrovirus."

The Great philosophers must be rolling over in their graves.

All the ideas of why we humans have ideas

All the fighting in the world because we have different

ideas

All the causes everyone in the world are fighting for

All the ideas that every one in the world believes make

them different from one another

All the people of this planet are here for one purpose

All of us are here to perpetuate the existents of an ancient

retro virus

All and every human on this planet is filled we R N A an altered D N A

We carry what scientist call junk genes that they think have no function

The purpose of 98% of the genes we carry that have no function are to perpetuate the entity of the first ancient retro virus that infected the first animal that became the human animal.

All the people of this planet are a bag full of ancient m R N A's programmed to live programmed to die programmed to think.

Programmed to kill each other.

All we need to do, is understand the R N A's program

Change it and we can all be immortal.

What would this planet be like if every human did not grow old

Did not go to war with each other

Because we know that all the people on this planet are a

host for the benefit of an ancient retro virus.

If we could fine the codes in our genetic system that

turns off the grow hormone in every male human at around

forty five to fifty years old that creates old age and disease

If we could find the codes that cause Menopause in

women at forty to fifty years old the cause old age and

disease

Yes all humans of this planet would be immortal.

The planet would be over populated

There would be no old people

No one would die

We have not found the codes that create old age and death.

We know they exist and we know what they do and we

know why.

We know that levels of Testosterone diminish in human

males

Normal Testosterone and free Testosterone in the body

Builds muscles and bones

Improves your immune system with the production of red

blood cells

Gives you energy mental concentration.

Improves your sex drive

With out Testosterone all of the above disappears

Disease of the bones and entire body is assessable due to a

weak immune system

And the journey to death begins

Menopause

We know that in women the decline of hormones creates all

of the physical problems and symptoms

Women, like men are susceptible to all the disease that men

acquire

Today there is no way to stop this because we don't

know the codes that create it but we can slow it down

I do not recommend Hormone Replacement Therapy

For men or women

For women in Menopause take Vitex and Black

Cohosh to balance the hormone problem

For men to release free Testosterone in the body

Take L-Ornithine and L-Arginine before bed no food no

protein with water

For men if you have the herpes' virus you should not take

L-Arginine you may try taking L-Lysine 1,000 mg with L-

Arginine and L-Ornithine the L-Lysine will inhibit

For women and men for a more information on how you

can live longer and healthier

Read my books

Vivre A Jamais

The Miracle of Anti Ageing

Vivre A Jamais

What they're not telling you about cancer cures

Vivre A Jamais

How to get a beautiful body and complexion using Ancient

herbs

Last night a government order came

To enlist boys who had reached eighteen.

They must help defend the capital…

O Mother! O Children, do not weep so!

Shedding such tears will injure you.

When tears stop flowing then bones come through,

Nor Heaven nor Earth has compassion then…

Do you know that in Shantung there are two hundred counties

turned to desert forlorn,

Thousands of villages, farms, covered only with bushes, the

thorn?

Men are slain like dogs; women driven like hens along.

If I had only known how bad is the fate of boys,

I would have had my children all girls.

Boys are only born to be buried beneath tall grass.

Still the bones of the war-dead of long ago are beside the blue

sea when you pass.

They are wildly white, and they lay exposed on the sand.

Both the little young ghosts and the old ghosts gather here to cry

in a band.

When the rains sweep down, and the autumn and winds chill,

Their voices are loud, so loud that I learn how grief can kill.

Birds make love in their dreams while they drift on the tide.

For dusk's path the fireflies must make their own light.

Why should man kill man in order to live?

In vain I sigh in the passing night.

www.ingramcontent.com/pod-product-compliance
Lightning Source LLC
Chambersburg PA
CBHW050907180526
45159CB00007B/2825